Knowledge's Kaleidoscopic Cartography

Visualizing information's diverse forms through interdisciplinary lenses

Clara Hendricks

Table of Contents

1 Introduction: Charting the Course

Data exists in countless forms, not just numbers in spreadsheets. Consider the nuances of a historical account, the intricate web of social interactions within a community, or the complex patterns revealed in a scientific experiment. Each represents data, but in vastly different ways. Recognizing this diversity is crucial; it's the first step towards mastering data visualization. Understanding that data is not monolithic, but multifaceted, is fundamental to developing effective visual representations. We need to grasp this heterogeneity to choose the right visual approach for each dataset. Failing to acknowledge this will lead to misinterpretations and ineffective communication. This initial understanding will serve as a solid base for our journey.

Visual literacy is more than just "seeing." It's about actively interpreting and understanding visual information. It involves critically analyzing what you see, identifying patterns, and drawing inferences. This isn't an innate ability; it's a skill honed through practice and exposure. Think of a complex map – you wouldn't instinctively understand its subtleties without prior experience with map reading. Similarly, we need to train our eyes to decode visual representations of data. This book aims to equip you with that skill set, enabling you to move beyond superficial observation towards insightful analysis. Sharpening

your visual literacy will allow you to become a more critical and effective data interpreter.
To effectively communicate data, visualization must be chosen carefully. A bar chart might perfectly illustrate sales figures across different regions, but it would be less suitable for depicting complex relationships between variables. Understanding your data's nature—its type (quantitative or qualitative), its volume, and the story you want to tell—directs your choice of visualization. Choosing the right tools, whether they be simple spreadsheets or sophisticated software packages, is crucial. This process involves deep consideration, moving beyond merely presenting information to fostering a compelling narrative. Therefore, selecting appropriate visualization methods aligns with your communication objectives.
Consider the difference between showing raw numbers and presenting them visually. A table full of figures might be accurate, but it's often overwhelming and difficult to interpret quickly. A well-designed graph, however, instantly reveals trends, patterns, and outliers. This is the power of visualization—transforming complex information into easily digestible insights. It's about making data accessible and engaging for your audience. Data visualization should not be a barrier, but a bridge to understanding. Its strength lies in its ability to transform the dense into the decipherable.
Effective visualization isn't just about aesthetics; it's about clarity and accuracy. A visually appealing chart that misrepresents data is worse than no chart at all. Maintaining ethical practices, avoiding biased representations, and accurately reflecting the data are paramount. This demands a meticulous approach, ensuring that your visualization accurately conveys the story your data is telling. Ethical considerations should guide every step, from data collection to final visualization, guaranteeing responsible use and reliable communication. The integrity of your visualization directly impacts the trust placed in your findings.
Finally, remember that data visualization is a continuous learn-

ing process. The field is constantly evolving, with new techniques and tools emerging regularly. Embracing this dynamism, staying updated with current practices, and engaging in continuous learning is essential to maintaining proficiency and becoming a skilled practitioner. The mastery of data visualization is a journey, not a destination, necessitating a commitment to lifelong learning and adaptation. Therefore, remain curious, seek new knowledge, and embrace the ever-changing landscape of data visualization.

1.1 Data's Diverse Forms

Data exists in a bewildering array of forms, far beyond simple numbers in spreadsheets. Understanding these diverse forms is the first step towards effective data visualization. Consider, for example, the seemingly straightforward numerical data from a scientific experiment: precise measurements of temperature, pressure, and reaction time. But this raw data, while quantifiable, needs context. Are these measurements taken in Celsius or Fahrenheit? Over what time period? Understanding the units, the sampling method, and any potential errors in the measurement process is as crucial as the numbers themselves. This contextual information, often non-numerical, shapes the meaning and interpretation of the quantitative data.
We also encounter qualitative data which, while not directly measurable in numerical terms, is equally rich in information. Think of customer reviews for a new product: positive comments, negative feedback, and neutral assessments all paint a picture of customer perception. These aren't numbers, but they hold immense value. Analyzing the frequency of certain words or phrases, identifying emotional tones, and categorizing feedback into themes reveal patterns and trends often invisible in raw text. These qualitative datasets—text, images, videos—demand different visualization strategies than numerical data. Interpret-

ing these requires nuanced understanding and careful consideration of the methods used to collect, process, and represent the qualitative information.

Another frequently overlooked form of data resides in our interactions. Think of the clicks, scrolls, and searches that define online behavior. These digital footprints leave a trail of information about user preferences, engagement levels, and website navigation. This behavioral data is frequently used in web analytics to understand user experience and optimize website design. Similarly, social media interactions—likes, shares, comments, and follower counts—provide insights into social networks, trends, and public sentiment. This data, however, can be complex to visualize. Visualizing this sort of relational data often involves network diagrams or interactive dashboards, showcasing connections and flows that might otherwise remain hidden.

Then there's the world of geospatial data, which blends location with other information. A simple map might show the distribution of a particular species of plant, but overlaying population density, elevation, or climate data transforms it into a powerful tool for analysis. This combined dataset paints a much more complete picture than each data point alone. Geospatial data visualization allows us to understand spatial correlations and distributions, leading to insightful conclusions that were impossible to see from individual sets of data alone. Similarly, temporal data, showing changes over time, demands visualization that accurately conveys the rate and extent of change. Line graphs, for instance, elegantly show trends over years, months, or even seconds, making the passage of time and the changes within it strikingly clear. These time-series visualizations reveal patterns and dynamics which are more easily understood through this visual approach.

Beyond these basic categories, we find more abstract data forms. These might include sensor data from smart devices, which constantly generate streams of information about temperature, light, motion, and other environmental factors. This

high-volume, high-velocity data needs sophisticated visualization techniques to make it comprehensible. In this context, we need real-time data visualizations – updating dynamically to capture the live flow of information. Furthermore, the data might have no inherent structure: think of unstructured text documents or images. Here, the challenge is transforming these less organized forms into visual representations that allow for insightful analysis and pattern identification. Techniques like word clouds, image similarity mapping, or thematic network diagrams can help reveal the underlying structure within this initially disorganized data.
Ultimately, understanding the diverse forms data can take is fundamental to effective data visualization. The choice of visualization technique is directly tied to the nature of the data itself—whether numerical, qualitative, temporal, geospatial, or some combination of these. By recognizing the nuances of different data types and employing appropriate visualization techniques, we unlock the power of data to inform, persuade, and enlighten. This forms the backbone of effective data fluency, enabling us to use visual tools to communicate effectively, make data-driven decisions, and leverage the wealth of information available to us in the increasingly data-rich world we inhabit. Mastering this ability is key to navigating the complexities of a rapidly changing information landscape and unlocks the transformative potential of effective communication in your chosen field.

1.2 Visual Literacy Basics

Visual literacy isn't just about seeing; it's about understanding what you see, interpreting the message conveyed, and using that understanding to make informed decisions. Think of a map – it's a visual representation of geographical information. Without visual literacy, a map is just a collection of lines and symbols.

With it, it's a tool for navigation and understanding spatial relationships. This understanding extends far beyond simple maps, encompassing charts, graphs, images, and even the layout of a website. Mastering this skill is crucial for navigating the deluge of visual information we encounter daily.

Developing your visual literacy involves honing several key skills. First, you must learn to recognize different visual forms and understand their typical conventions. A pie chart represents proportions; a bar chart compares quantities; a line graph shows trends over time. Recognizing these forms allows you to quickly grasp the data's core message. Beyond these basic forms, there are countless variations, each with subtle differences that can significantly alter the conveyed information. The ability to decipher these nuances is paramount for accurate interpretation. Understanding these variations requires dedicated study and practice, a journey to becoming visually fluent.

Beyond recognizing visual forms, critical evaluation is crucial. Consider the source of the visual; its creator's biases might influence the data's presentation. Look for misleading scales, truncated axes, or selective data representation. These techniques, often employed intentionally or unintentionally, can subtly manipulate the viewer's interpretation. Effective visual literacy demands a skeptical, analytical eye, constantly questioning the message being presented. Always strive for balance; challenge your own assumptions as well. Question the visual, not just the data.

Furthermore, effective visual literacy involves understanding the context in which a visual is presented. The same chart might have vastly different meanings depending on the surrounding text, the accompanying caption, or the overall article's aim. For instance, a chart showing an increase in crime rates might be interpreted differently if presented in an article about increased police funding versus an article highlighting social inequalities. Context provides crucial clues that shape our understanding. Ignoring it leads to misinterpretations and inaccurate conclusions.

Never look at data in isolation.
Next, consider the intended audience. A complex data visualization suitable for a scientific journal might be completely unintelligible to a lay audience. Visual literacy includes adapting visual communication techniques to reach different audiences. Effective communicators use visual language that resonates with their audience, tailoring complexity to ensure the message is clearly understood, no matter the background of the person observing.
Creating your own visualizations is another aspect of visual literacy. The process of designing a visual forces you to think critically about the data, identify key patterns, and select the most effective method to represent your findings. You need to consider what you want your audience to understand, and then plan out the most effective presentation to convey that understanding. By actively creating visualizations, you solidify your understanding of the principles discussed and refine your visual reasoning skills. It's not just about making a pretty picture; it is about strategic communication through visuals.
Finally, understand that visual literacy is an evolving skill. New visualization techniques and technologies are constantly emerging. Staying abreast of these developments and expanding your knowledge base is crucial for maintaining a high level of visual fluency. Explore new platforms, experiment with different software, and regularly challenge yourself with novel visual forms. In this constantly evolving world, adaptation and learning are paramount in maintaining visual fluency. Embrace change; it is an essential ingredient in mastering visual literacy. The ability to adapt and apply these principles will empower you in navigating the world of visual information.

2 Visualizing Quantitative Data

Bar charts, line graphs, and scatter plots are fundamental tools; mastering them unlocks a world of insightful visualizations. A simple bar chart can powerfully compare categories, revealing immediate differences in quantities. Line graphs, on the other hand, excel at showcasing trends over time, elegantly illustrating growth, decline, or cyclical patterns. Scatter plots, with their ability to display the relationship between two variables, identify correlations and outliers, hinting at underlying patterns that might otherwise remain hidden. The choice depends entirely on the nature of your data and the story you want to tell. Remember, a well-chosen chart isn't just about presenting data; it's about communicating a clear, concise narrative.

Understanding the nuances of each graph type is critical. For instance, while a bar chart effectively compares discrete categories, using it for continuous data can lead to misinterpretations. Similarly, a line graph assumes a continuous relationship between data points, making it unsuitable for representing independent categories. Pay close attention to your axes, ensuring they are clearly labeled and scaled appropriately to avoid distortion or misrepresentation. Accurate labeling and scaling are the cornerstones of ethical and effective data visualization. Think of it as building a bridge; the foundation, like accurate labeling, must

be solid to ensure a safe crossing.
Statistical mapping transforms raw numerical data into visually compelling geographical representations. Imagine mapping crime rates across a city, using color intensity to represent the severity of the problem. Such a map instantly reveals crime hotspots, providing valuable insights for law enforcement and urban planning. This technique extends beyond crime; it finds applications in analyzing demographic data, tracking disease outbreaks, visualizing economic indicators, or exploring environmental impact across various regions. The key is choosing the appropriate mapping technique to suit the nature of your data and your analytical goals. Choropleth maps, for instance, excel in depicting variations across regions using different color shades.
The availability of various software and online tools significantly simplifies the process of data visualization. Programs like Excel offer basic charting capabilities, while more sophisticated tools like Tableau or R provide advanced analytical features and a wider range of visualization options. These tools allow for interactive exploration of datasets, enabling users to dynamically manipulate visualizations and filter data to gain specific insights. Online resources, including tutorials and documentation, are readily available, providing support for beginners and experienced users alike. Experiment with various options; find the tools that suit your skill level and the complexity of your data. The perfect tool is the one that empowers you to communicate your findings effectively.
Beyond the basics, consider adding contextual information to enhance your visualizations. Adding clear titles and concise captions provides essential context, guiding the viewer's understanding and interpretation of the data. A well-written caption can be the difference between a compelling visualization and a confusing display. Think about your audience; tailor your visuals and captions to resonate with their existing knowledge and understanding. A visualization should not only present

data; it should effectively communicate your findings to the intended recipient. Furthermore, consider integrating your visualizations into narratives, strengthening the impact of your message. Weaving data into a story makes it more relatable and engaging. Always remember your goal: clear communication and insightful understanding.

Visualizing quantitative data isn't merely about creating charts and maps; it's about crafting compelling narratives. The process requires careful consideration of the data's nature, the chosen visualization technique, and, most importantly, the audience's understanding. Therefore, refining your communication skills becomes as crucial as mastering software tools. Effective communication ensures that your insights are not only accurately presented but also readily grasped and appreciated by your intended audience. After all, even the most intricate visualization is useless if it fails to connect with its intended recipient. Effective data visualization is a blend of technical proficiency and clear communication.

The power of visualizing quantitative data lies in its ability to transform complex numerical information into easily digestible visuals. From identifying trends to revealing hidden patterns, these visualizations provide valuable insights into a wide range of topics. However, remember that the effectiveness of any visualization depends not just on its technical accuracy but also on its clarity and ability to convey meaning effectively. A poorly designed visualization, even if technically accurate, can be more misleading than helpful. Always strive for clarity, accuracy, and effective communication. These are the keys to unlocking the full potential of data visualization. Your goal is not just to show the data; it's to reveal the story it holds.

Finally, remember that data visualization is an iterative process. You'll likely refine your visualizations multiple times as you gain deeper insights and better understand your audience. Don't be afraid to experiment with different approaches, and don't hesitate to seek feedback from others. Effective data visualization is a journey, not a destination. As you improve your skills, you'll develop a more refined approach to creating effective and insightful visualizations. Always aim for improvement; the ability to visualize data effectively is a skill constantly refined with practice and experience. The journey towards mastering data visualization is a continuous learning process, one you will constantly refine as you progress.

2.1 Graphs & Charts

Bar charts, perhaps the most familiar type of graph, excel at comparing different categories. Their simplicity makes them incredibly accessible, ideal for conveying straightforward comparisons of quantities. Think of sales figures across different regions, or the popularity of various products within a specific market. The visual height of each bar instantly communicates the magnitude of each category, requiring minimal cognitive effort from the viewer. This directness makes them a powerful tool for presentations and reports aimed at a broad audience, fostering quick and clear understanding. Remember to label axes clearly and choose appropriate scales to avoid misleading interpretations. Using consistent colors and spacing further enhances readability and impact.
Line graphs, on the other hand, shine when displaying trends over time. Think of stock prices fluctuating over months, or temperature changes throughout a year. The continuous line visually connects data points, revealing patterns and fluctuations that might be missed in other visual representations. This dynamic nature makes them especially useful for demonstrating growth, decline, or cyclical patterns. Careful consideration of the timescale is crucial; choosing a scale that accurately reflects the data's nuances is key to preventing a distorted or misleading representation of the trend. For instance, a short time frame may exaggerate small changes, while an overly large time frame

might obscure significant shifts.
Scatter plots offer a different perspective, ideal for exploring correlations between two variables. Imagine plotting student test scores against the number of hours they studied. Each point represents a single data point, its position revealing the relationship between the two variables. Clustering of points might indicate a strong positive or negative correlation, while a scattered distribution suggests a weak or nonexistent relationship. Interpreting scatter plots involves looking for trends and patterns, taking note of outliers that might warrant further investigation. Remember to label axes clearly and consider using different colors or shapes to represent distinct subgroups within your data. A clear legend is essential when interpreting complex scatter plots with multiple subgroups.
Pie charts effectively show the proportion of parts to a whole. Consider a budget breakdown, showing the percentage allocated to different expenses like rent, food, and entertainment. Each slice visually represents a portion, its size directly reflecting its proportional contribution to the whole. Their intuitive nature makes them easily understandable, particularly for audiences with limited data visualization experience. However, overuse can lead to cluttered visuals, particularly when dealing with numerous categories. Keeping the number of categories manageable and labeling each slice clearly are essential for maintaining clarity and avoiding visual overload.
Histograms provide insights into the distribution of data. Imagine plotting the heights of students in a class. Unlike bar charts, histogram bars represent ranges of values, revealing the frequency of data points within each range. The shape of the histogram reveals important characteristics of the data's distribution, such as its symmetry, skewness, and the presence of outliers. This can be particularly valuable for understanding central tendencies like mean, median, and mode, offering a richer understanding than simple summary statistics. Choosing appropriate bin sizes is crucial, as overly large bins can obscure

important details, while overly small bins can lead to a noisy and difficult-to-interpret histogram.
Beyond these fundamental chart types, numerous specialized charts exist for specific purposes. Area charts, for instance, are variations of line graphs, useful for visualizing cumulative data. Box plots offer a concise summary of data distribution, showing median, quartiles, and outliers. Heatmaps display data as colors, making it easy to spot patterns and anomalies in large datasets. Choosing the right chart type depends on the type of data you are analyzing and the message you wish to convey. Careful consideration of your audience and the specific insights you want to highlight should guide your selection process. Always remember, the goal is clear communication; a well-chosen chart enhances understanding, while a poorly chosen one obscures it. The most effective visualizations are clear, concise, and easily interpreted. They avoid unnecessary complexity and present the data in a manner that directly supports the intended narrative.

2.2 Statistical Mapping

Statistical maps transform raw data into compelling visual narratives, revealing patterns and relationships geographically. Imagine a heatmap showing crime rates across a city; instantly, hotspots emerge, guiding policing strategies and resource allocation. This is the power of statistical mapping—transforming complex data sets into easily digestible visual representations. It's a crucial tool for understanding spatial distributions and identifying trends, allowing us to analyze everything from population density to disease outbreaks. Effective use of this technique depends on careful data selection and an understanding of the inherent limitations.
Understanding your data is paramount before even considering a map. Are your data points accurately located? What is the

spatial resolution of your data? Are there significant gaps or biases in your data collection? Addressing these questions is critical to creating a trustworthy and informative map. Poor data leads to misleading visuals, undermining the entire purpose of the exercise. Consider the context of your data – a map illustrating income distribution requires different considerations than one showing earthquake epicenters. The level of detail, the choice of color scheme, and even the map projection can significantly influence the interpretation of the data.

Choosing the right type of map is essential to communicating your data effectively. Dot density maps are excellent for illustrating the concentration of points, showcasing population density, for example. Choropleth maps use color shading to represent data values across geographic areas; these are ideal for visualizing variables like income levels or voting patterns across counties or states. Isoline maps, on the other hand, connect points of equal value, creating contour lines perfect for representing elevation or temperature gradients. Each map type serves a unique purpose, and selecting the most appropriate one depends entirely on the nature of your data and the message you're trying to convey. Failing to match the map type to your data will lead to confusion and misinterpretations.

The selection of a color scheme isn't just an aesthetic choice; it significantly impacts the clarity and interpretation of your map. A well-chosen palette enhances understanding, while a poorly chosen one can lead to misinterpretations. Consider using a sequential color scheme for representing data that progresses continuously, like temperature or elevation. Diverging schemes are best for showing data that varies around a central value, such as temperature differences from an average. And categorical schemes are best for differentiating distinct groups, like land-use types. Furthermore, avoid using too many colors, as this can overwhelm the viewer and hinder their ability to grasp the key patterns. Accessibility is another vital factor; ensure your color scheme considers users with color blindness.

Beyond the technical aspects, ethical considerations are crucial. Statistical maps can be easily manipulated to present a biased or misleading narrative. This is especially true when dealing with sensitive data, such as income inequality or crime rates. Transparent data handling, careful selection of visualization methods, and clear labeling are essential to avoiding such pitfalls. Always disclose any limitations or potential biases in your data, and strive for an objective presentation that allows for critical interpretation by the audience. Transparency in data sources and methodology is key to establishing trust and ensuring the integrity of your visual representation.

Software plays a critical role in creating effective statistical maps. Geographic Information Systems (GIS) software packages, such as ArcGIS or QGIS, offer powerful tools for data manipulation, spatial analysis, and map creation. These tools allow you to integrate various data sources, perform complex analyses, and create visually appealing and informative maps. Numerous online mapping tools are also available, providing more accessible options for users with less technical expertise. However, mastering the functionality of even basic software requires practice and understanding of fundamental concepts. The more complex the data or the specific output goals, the more relevant it becomes to develop specific skills in this area.

Finally, remember the audience. Your map should be clear, concise, and easily understood by its intended audience. Use clear labels, a legend that is easy to follow, and a title that accurately reflects the data being presented. Avoid unnecessary clutter, and focus on highlighting the key patterns and relationships within the data. A well-designed statistical map provides insights, prompts questions, and tells a compelling story, enabling data-driven decision-making. Therefore, crafting the map well is an act of effective communication, just as much as of data analysis.

2.3 Data Visualization Tools

Choosing the right tool is crucial for effective data visualization. Many options cater to different needs and skill levels, from simple spreadsheet software to advanced programming libraries. Microsoft Excel, while seemingly basic, remains surprisingly powerful for creating charts and graphs, particularly for beginners. Its user-friendly interface makes it accessible, allowing you to quickly visualize simple datasets with common chart types like bar graphs, pie charts, and line graphs. However, its limitations become apparent with larger, more complex datasets, or when more sophisticated visualization techniques are required. Remember, the effectiveness of your visualization hinges on choosing the tool that best matches your data and the insights you aim to convey.

Moving beyond spreadsheets, consider dedicated data visualization software. Tableau and Power BI are industry leaders, offering robust features for data cleaning, transformation, and interactive visualization. These platforms handle large datasets efficiently and provide a wide range of chart types and customizable options. They excel at creating interactive dashboards, allowing users to explore data dynamically and gain deeper insights. Learning curves exist for both, demanding time investment to master their full capabilities. Nevertheless, the resulting visualizations are often significantly more polished and insightful than those achievable in simpler tools. Their strengths lie in the creation of sophisticated, interactive visualizations ideal for presentations and reports.

For those comfortable with coding, programming languages like Python and R offer unparalleled flexibility and control. Libraries such as Matplotlib, Seaborn, and Plotly in Python, and ggplot2 in R, provide extensive functionalities for creating custom visualizations tailored to your specific requirements. The learning curve is steeper, demanding programming knowledge and familiarity with the chosen libraries' syntax and functionalities. This investment, however, grants access to a vast range of visualization techniques and the ability to automate the visualization process for large-scale projects. This level of control is invaluable for researchers and data scientists who need highly tailored and custom visualizations.

Beyond dedicated software and programming, online tools provide accessible and often free solutions. Many web-based platforms offer intuitive drag-and-drop interfaces for creating basic visualizations without requiring any coding knowledge. These tools are ideally suited for quick visualizations of smaller datasets and for users without prior experience in data analysis or software programming. However, limitations in customization and scalability are often inherent. Consequently, these platforms are most suitable for simple, quick visualizations rather than complex, detailed analysis.

Ultimately, the best tool depends on your specific needs and technical skills. Beginners might find Excel sufficient for simple visualizations, while experienced users may benefit from the power and flexibility of Python or dedicated software like Tableau. Explore different options, experiment with different tools, and choose the one that best supports your workflow and allows you to effectively communicate your data insights. The key is to find a balance between ease of use and the ability to create visualizations that are both informative and engaging to your audience, remembering that the goal isn't merely to display data, but to effectively communicate its underlying story. Consider the type of data you are working with and tailor the choice of tool to achieve maximum impact and clarity in your visualizations.

3 Qualitative Data Visualization

Network analysis offers a powerful way to visualize relationships within qualitative data. Imagine mapping the connections between characters in a novel, revealing intricate webs of alliances and conflicts. Or consider using it to understand the spread of information within a community, highlighting key influencers and communication channels. The visual representation, whether a simple node-link diagram or a more complex network graph, unveils patterns and insights hidden within the textual or observational data. Software packages like Gephi and NodeXL can assist in creating these visualizations, allowing for interactive exploration and analysis. Remember to carefully consider the type of relationships you are analyzing and choose appropriate visualization techniques to best represent those connections. The visual clarity of these networks facilitates understanding complex systems and identifying influential actors, a crucial element for researchers across disciplines. Effective use requires clear definition of nodes and edges, ensuring the visualization accurately reflects the underlying data. Proper annotation and interactive features can enhance interpretation and understanding. The impact of different network layouts on the perception of relationships should also be carefully considered. For instance, a force-directed layout can effectively represent complex relationships where a hierarchical layout would be less appropriate.

Concept mapping provides a visual structure for organizing and representing ideas. Unlike network analysis that focuses on relationships, concept mapping emphasizes hierarchical structures and the connections between concepts. It's a valuable tool for brainstorming, knowledge synthesis, and communicating complex ideas in a concise, easy-to-understand manner. Start by identifying the central concept and then branch out to related ideas, creating a hierarchical structure. Use linking words or phrases to show the relationships between concepts. Consider using different shapes, colors, or sizes to highlight key concepts or emphasize relationships. Many software tools can assist in creating and managing concept maps, such as MindManager or FreeMind. For maximum impact, your concept map needs to be clear, concise, and logically structured. Overly complex maps can be difficult to understand and might obscure your message rather than clarify it. Careful consideration of visual hierarchy and the use of whitespace helps in enhancing readability and comprehension. By employing consistent visual cues, your concept map effectively translates complex ideas into a readily graspable format.

Narrative visualization goes beyond simply presenting data; it aims to tell a story using visuals. This approach combines data with narrative elements, creating engaging and memorable visualizations. Begin by identifying a central narrative thread, a story you want to tell using your data. Then, choose appropriate visual elements to represent the key events, characters, or themes within that narrative. Consider using timelines, maps, or charts, combined with images, text, and other visual elements to make the story compelling. The goal is to engage the audience emotionally while informing them. This technique excels in conveying complex processes or personal experiences, making data more accessible and relatable. Think of how a timeline can showcase the progression of a historical event, using images and text to bring the narrative to life. Or consider how a map might visually represent the journey of a migrant, showing their path and struggles along the way. Effective storytelling ensures that your visualization isn't just informative, but also evocative and memorable. Careful consideration of sequencing and pacing is critical to maintain audience engagement. The most effective narrative visualizations use strong visual metaphors to create a lasting impression.

Infographics are concise, visually appealing representations of information. They are designed for quick comprehension and are widely used for communicating data in a variety of contexts. To create an effective infographic, start by identifying your key message and the data you need to support it. Then, choose a visual style and layout that fits your message and target audience. Keep it concise and easy to read, using clear headings, labels, and visuals. Infographics thrive on simplicity; avoid overwhelming the viewer with too much information. The use of strong visuals and a clear, concise writing style is crucial. Consider using icons, charts, and other visual elements to present data in an easily digestible format. Choose a visually appealing color palette and maintain a consistent visual style throughout the infographic. Tools such as Canva or Piktochart can simplify the design process. Remember the purpose: an infographic's success hinges on its ability to convey complex information quickly and memorably.

Visual ethnography uses visual methods to represent cultural insights and understandings. Unlike the previous methods focusing on data relationships, visual ethnography focuses on capturing the nuances of human experience and culture. It involves using photographs, videos, drawings, maps, and other visual media to capture and communicate observations, interpretations, and findings. The focus is on exploring and depicting lived experience. Ethical considerations are paramount; always obtain informed consent and be mindful of cultural sensitivity. A key aspect is using visuals to tell stories and convey meanings that might otherwise be missed using traditional textual reporting methods. Consider creating photo essays, videos, or interactive maps to showcase the rich detail of cultural practices and beliefs. Careful analysis of the chosen media and their framing is essential to ensure accurate representation and avoid misinterpretations. By skillfully incorporating visual and textual elements, visual ethnography constructs comprehensive and insightful narratives that surpass the limitations of textual descriptions alone.

3.1 Network Analysis

Network analysis is a powerful technique for visualizing relationships between entities. Imagine a social network, where individuals are nodes and connections represent friendships. This visual representation immediately reveals clusters of friends, influential individuals (those with many connections), and perhaps even isolated groups. This simple example showcases the core strength of network analysis: uncovering hidden patterns in complex relationships, allowing for a deeper understanding of the system's structure and dynamics. The beauty lies not just in the visualization itself, but the insights gained – understanding how information flows, identifying key players, and predicting future behaviors.

Beyond social networks, network analysis finds applications across diverse fields. Consider a citation network in academic research, where papers are nodes and citations are the connecting links. Visualizing this network reveals influential papers, research communities, and the evolution of ideas over time. In epidemiology, networks can map the spread of infectious diseases, showing how infections jump from person to person, allowing for more efficient control measures. Business applications include analyzing supply chains, identifying key partners, and understanding the flow of goods and resources. This adaptability makes network analysis an indispensable tool for data

interpretation and communication across many disciplines.
Creating a network visualization begins with defining the nodes and edges. Nodes represent the individual entities within your dataset—these could be people, documents, locations, or any other discrete unit. Edges, on the other hand, represent the relationships between those nodes—interactions, influences, or connections. The type of edge you choose depends heavily on the data; it could be a simple binary connection (connected or not), or it could represent a weighted relationship (strength of the tie, frequency of interaction, etc.). Careful consideration of these choices is crucial for generating a meaningful and accurate representation. A poorly chosen representation can lead to misleading conclusions.
The visualization itself can take several forms. A simple, unweighted network can be visualized as a graph with nodes and lines connecting them. This makes it easier to spot clusters, isolated nodes, and central figures. For weighted networks, you might use thicker lines to represent stronger connections, or you could use color to encode additional information about the edges. For larger, more complex networks, more sophisticated layout algorithms are necessary to prevent the visualization from becoming an uninterpretable mess. Software such as Gephi or Cytoscape provides powerful tools for creating and manipulating these visualizations. It's important to select the appropriate tool depending on the size and complexity of your data.
Network analysis provides not only visual insights but also a wealth of quantitative data. Metrics such as degree centrality (the number of connections a node has), betweenness centrality (the number of shortest paths between other nodes that pass through a given node), and closeness centrality (the average distance from a node to all other nodes) offer ways to quantify influence, connectivity, and information flow within the network. These metrics, coupled with the visual representation, can be instrumental in testing hypotheses, revealing hidden structures, and confirming or refuting your findings. Understanding and

utilizing these metrics is key to drawing meaningful interpretations from your analysis.

Data preparation for network analysis involves several crucial steps. First, the dataset needs to be cleaned and standardized. Inconsistent data entry and incomplete information are major hindrances for efficient network analysis. If your data is spread across several files, it must be integrated into a uniform format. Second, your data needs careful review for biases that may skew the resulting network visualization. Biases in the data collection, for example, might create inaccurate or incomplete representations of the relationships within the system. Data quality plays a significant role in the accuracy and reliability of the visualization and consequent inferences.

Interpreting network visualizations isn't just about identifying patterns; it also involves considering the limitations of the analysis. The visualization is a representation of the data you fed into it, and its accuracy depends on the quality and completeness of your data. The chosen algorithms for layout and centrality measures will also influence the appearance of the network and the values derived from the analysis. Always carefully consider the assumptions underpinning your approach, and communicate those assumptions clearly when presenting your findings to avoid potentially biased conclusions. This allows for honest evaluation of the results and their implications.

Finally, always remember the ethical considerations. Network analysis, particularly when dealing with sensitive data, requires responsible data handling and responsible visualization. Protecting the privacy of individuals whose data is being represented is of paramount importance. Anonymization techniques might be necessary to protect personal identities, particularly when representing social networks. Remember to be meticulous in your data management procedures to ensure confidentiality and adherence to privacy regulations. The potential benefits of this analysis must always be carefully weighed against the possible risks to individual privacy.

3.2 Concept Mapping

Concept mapping is a powerful technique for organizing and representing knowledge visually. It's particularly useful for understanding complex ideas, brainstorming, and planning projects. Unlike other visualization methods, concept mapping focuses on the relationships between concepts, not just the concepts themselves. Think of it as a visual brainstorming session, allowing you to explore connections you might miss using linear note-taking methods. This visual approach unlocks a new level of understanding, moving beyond simple lists and embracing the interconnectedness of information. You can use it to explore connections between your own thoughts and ideas, or you can adapt it to analyze concepts from other sources, such as research papers or lectures. This flexible nature of the technique means it will be useful throughout your academic journey, enabling you to make connections across different subjects and methodologies.

A key element of effective concept mapping lies in the careful selection of central concepts. These are the core ideas around which your map will be built. Identify the most important and relevant concepts related to your topic, ensuring that they are clearly defined and understandable. Once you've established your central concepts, you can start branching out, adding related ideas and connecting them with linking words or phrases. These connections are crucial, showing the relationships between concepts—whether they are hierarchical (one concept is a subcategory of another), sequential (one concept leads to another), or associative (concepts are related in some other way). Don't be afraid to experiment with different types of connections; exploring these relationships helps you deepen your understanding of the topic at hand.

The visual representation is key to concept mapping. You'll use nodes, or circles or boxes, to represent your concepts. These nodes are linked together by lines or arrows representing the connections between concepts. Simple symbols or short descriptive phrases can further enhance the clarity of the connections between the concepts. Don't overcomplicate the process; the goal is clarity and understanding. The visual aspect allows for an overview of the relationship between concepts. This visual approach is significantly different from the linear method of note-taking and allows for a more holistic approach to understanding the concept. You are not merely recording; you are actively building understanding through visual representation.

Creating a concept map is an iterative process. Rarely will your first attempt be perfect. You'll likely find yourself adding, removing, or re-arranging concepts as your understanding deepens. This dynamic nature of concept mapping is a strength. It reflects the non-linear process of learning and understanding. Remember, it's a tool for your thinking, not a finished product to be judged on aesthetic appeal. The evolution of your map as your understanding evolves is a reflection of effective engagement with the material. The benefit comes in the process, not the perfection of the end result.

Concept maps have practical applications across numerous fields. Researchers use them to outline their studies, showing the relationships between variables and hypotheses. Students use concept maps as a study aid, consolidating information from various sources, and organizing it into a coherent structure. Even in business settings, they can be used for project planning and problem-solving, illustrating the steps involved and the dependencies between tasks. You can use them to summarise complex readings and present your findings in a concise, clear manner. The versatility of concept mapping makes it a truly valuable tool for understanding and expressing complex ideas in a manner that is accessible and visually compelling, transforming abstract ideas into a concrete and manageable structure. The key lies in its flexibility and adaptability to diverse contexts and learning styles.

The process might seem daunting initially, but with practice, it becomes second nature. Start with a central concept, branch out with related ideas, and refine your map as your understanding grows. Don't be afraid to experiment with different layouts and visual cues to see what works best for your thinking style. Remember that the most valuable part is in the active process of connecting concepts, uncovering connections, and deepening your understanding through this interactive approach. By repeatedly engaging with this process, you'll find it easier to form connections between your thoughts, ideas, and concepts, leading to stronger comprehension, and better retention and application of the knowledge acquired. Concept mapping is not just a technique but a powerful tool for thinking differently and effectively visualizing knowledge.

3.3 Narrative Visualization

Narrative visualization isn't merely presenting data; it's crafting a compelling story. Think of it as weaving a tapestry where each thread is a data point, contributing to a larger, more meaningful picture. Unlike simple charts and graphs, narrative visualization prioritizes the narrative arc, using visuals to guide the reader on a journey of understanding. This approach requires careful consideration of the audience and the desired message. You aren't simply displaying facts; you're constructing an experience.

Effective narrative visualization starts with a clear storyline. Identify the key message you want to convey. What's the central argument or insight your data reveals? Organize your data to support this narrative. Consider the sequence of events or ideas, building towards a climax or a resolution. This process may involve identifying pivotal moments, key characters (if applicable), and the relationships between them all, ensuring a logical flow of information that captivates the reader's attention. This structure will form the backbone of your visualization. Without a strong narrative, even the most visually stunning representation will fail to resonate.

Choosing the right visual elements is paramount. Images, icons, and illustrations can enrich the narrative, adding depth and emotional resonance. For instance, a photograph of a smiling child might complement data illustrating improved healthcare outcomes. A map can effectively showcase geographical distribution. However, avoid overwhelming the reader with too many visuals; prioritize clarity and relevance, ensuring each element contributes meaningfully to the unfolding story. Every single choice should enhance the narrative's impact and improve the overall understanding and the storytelling effectiveness.

The power of narrative visualization lies in its ability to humanize data. Abstractions become tangible; numbers transform into stories of people's lives. Consider incorporating personal anecdotes or case studies to illustrate trends or patterns. This personal touch can make complex information more accessible and relatable. Imagine supplementing crime statistics with the story of a victim's family. Or, showing statistics about climate change alongside pictures of people whose homes have been lost to floods or wildfires. Remember, connecting with the reader on an emotional level fosters deeper understanding and engagement. The goal is to make the data not just comprehensible, but also memorable and impactful.

Interactive elements can elevate the narrative further, allowing the reader to explore the data at their own pace. Think of timelines that allow users to zoom in on specific periods, or maps that let them drill down to individual locations. These interactive components transform a passive experience into an active engagement with the information. However, remember to maintain a clear narrative structure even with added interactivity, so the reader doesn't get lost in an overly complex system.

Beyond simple charts, a good narrative visualization embraces a range of visual techniques. Consider using timelines to illustrate historical trends. Employ maps to show geographic distributions. Integrate charts and graphs to support key points. Combine different visual elements to build a coherent and compelling narrative. Don't be afraid to experiment with different layouts and approaches to find the most effective way to tell your story. The combination of diverse visuals will give an amazing experience to your viewers.

Finally, remember the importance of context and accessibility. Provide clear labels, captions, and explanations to guide the reader through the narrative. Ensure the visualization is accessible to a wide audience, regardless of their technical expertise or familiarity with the subject matter. Use clear and concise language, and employ appropriate color schemes and visual cues for easy interpretation. Accessibility and clear communication are crucial for effective storytelling, therefore ensuring everyone can understand and benefit from the narrative you construct. Always put your audience first in the design process.

3.4 Infographics

Infographics condense complex information into easily digestible visual formats. They're not just pretty pictures; they're powerful communication tools, capable of conveying intricate data sets with clarity and impact, far surpassing the limitations of lengthy text reports. Think of them as visual storytelling, using images, charts, and minimal text to engage your audience and leave a lasting impression. Successful infographics require careful planning and execution; the key is to prioritize the most important information, making it stand out while avoiding clutter.

A well-designed infographic starts with a clear objective. What story are you trying to tell? What key insights do you want your audience to take away? Defining this beforehand is crucial; it dictates the structure, visual elements, and overall design. Without a defined goal, your infographic risks becoming a confusing jumble of unrelated information. Before even considering the visual design, spend time organizing your data, identifying the most compelling narratives, and structuring the information in a logical sequence. Consider your target audience—their level of understanding, interests, and preferred learning styles—and tailor the information and visual style accordingly. This isn't about cramming everything in; it's about presenting the most essential information effectively.

Choosing the right visual elements is paramount. While bar charts and pie charts are familiar choices for quantitative data, don't be afraid to explore other options. Maps can illustrate geographical distributions effectively, while icons and illustrations can add visual interest and clarify complex concepts. The key is to select visuals that accurately represent the data and complement the narrative. Maintain consistency in your visual style; using a coherent color palette and font choice will enhance readability and create a professional aesthetic. Aim for a balance between visual appeal and informational clarity; avoid excessive ornamentation that detracts from the main message. Remember, the goal is to communicate effectively, not to create a masterpiece of graphic design.

The layout of your infographic is also vital. Guide the reader's eye through the information logically using clear visual cues, like arrows, lines, or different color blocks. Break down the information into easily digestible chunks, using headings and subheadings to create a clear hierarchy. Ensure proper spacing to avoid a cluttered appearance. White space is your friend; it allows the reader's eye to rest and improves overall readability. Consider the flow of information; a linear approach may be suitable for a chronological narrative, while a more complex structure might be necessary for presenting interconnected concepts. Always keep your target audience in mind; their understanding of the subject matter will influence your design choices. The best infographics are intuitively understandable, requiring minimal effort from the reader to grasp the core message.

Finally, testing and iteration are crucial. Once you've completed a draft, share it with others for feedback. Are they able to easily understand the information? Does the infographic effectively communicate the intended message? Use this feedback to refine your design and make improvements. This iterative process is vital; it helps to identify any areas where clarity can be enhanced or the information presented more effectively. Don't be afraid to experiment with different layouts and visual elements until you've achieved a design that is both visually appealing and highly effective in communicating your information. Remember that the goal is clear, concise communication—a testament to the power of visual storytelling. A successful infographic leaves a lasting impact, making complex information memorable and easily understood.

3.5 Visual Ethnography

Visual ethnography offers a powerful lens through which to understand and represent cultural phenomena. Unlike traditional ethnographic methods that rely heavily on textual analysis, visual ethnography incorporates diverse visual media – photographs, videos, maps, even digitally created representations – to capture the richness and complexity of human experience within a specific cultural context. This approach allows for a more nuanced understanding of behaviours, interactions, and beliefs, moving beyond simple descriptions to reveal subtle patterns and meanings that might otherwise be missed. The integration of visual data significantly enriches qualitative analysis, enabling researchers to explore visual culture, social interactions, and community dynamics. It goes beyond simply observing; it's about interpreting visual data as a means of telling powerful stories and building empathy.
Consider, for instance, a study examining the impact of gentrification on a particular neighbourhood. Traditional methods

might rely on interviews and surveys, offering valuable information but potentially missing the nuanced visual cues present within the altered environment. Visual ethnography, however, would incorporate photographs of changing building facades, abandoned storefronts, community murals, and public spaces – offering a more comprehensive picture of the lived experience of residents. The visual record allows researchers to chart subtle changes in the physical landscape alongside the narrative accounts of residents, demonstrating the tangible effects of gentrification. This richer dataset provides a more persuasive and impactful narrative for any analysis, making it particularly valuable for community engagement and advocacy.

The process of conducting visual ethnographic research necessitates careful planning and ethical considerations. Gaining informed consent from participants is paramount. Transparency about the research aims and intended use of visual materials is crucial; researchers must be upfront about how the images will be used and the implications for participants' privacy. Researchers should build trust and rapport, fostering open communication and ensuring that participants feel comfortable and empowered throughout the research process. Ethical guidelines must be followed scrupulously, avoiding exploitation and ensuring that the representation of the culture is accurate and respectful. This proactive ethical approach can minimize harm and enhance the validity of the study.

Analyzing visual data within a visual ethnography study requires a different set of skills and methodologies than those used in traditional qualitative research. Simple observation is not enough; deep interpretive analysis is needed to understand the cultural significance of the images collected. This often requires collaboration with other researchers who may offer expertise in particular areas, such as semiotics, visual communication, or even photography itself. This interdisciplinary approach enriches the analysis and allows for a comprehensive interpretation of the visual data collected. For example, analyzing a collection of photographs requires careful consideration of framing, lighting, and composition – all of which can communicate implicit meanings. Similarly, analyzing video data necessitates careful attention to non-verbal cues, body language, and the spatial arrangement of individuals. Understanding these elements is crucial for drawing meaningful conclusions from the study.

The final stage of visual ethnography involves the representation and communication of findings. This can take many forms, from academic publications incorporating carefully selected images and video clips, to engaging exhibitions and interactive websites designed to share the research findings with broader audiences. The creation of visually compelling presentations of the research is crucial, ensuring that the complex interplay of visual data and analytical interpretations is effectively communicated. This stage emphasizes storytelling, using visuals to create an impactful narrative that speaks beyond the academic community. It's about translating the richness and complexity of cultural experiences into compelling and accessible formats. For example, a documentary film could integrate the visual data collected, alongside interviews and narration, to create a powerful narrative accessible to a wider audience. This method would offer a more engaging and accessible way to share the findings compared to a traditional academic paper. The ultimate goal is to foster understanding and empathy, furthering dialogue and potentially facilitating change.

4 Interdisciplinary Approaches

Historians, for example, can leverage data visualization to illuminate patterns and trends in historical events. Imagine charting the migration patterns of populations across centuries, visually representing the ebb and flow of empires, or mapping the spread of ideas and innovations. By using network analysis, historians can uncover hidden connections between individuals and events, enriching our understanding of the past. Interactive timelines and geographically-mapped data sets can allow for deeper exploration, encouraging a more dynamic and engaging approach to historical analysis than traditional textual methods. This allows students and researchers to interact directly with the data, fostering a more critical and informed understanding of historical narratives. Furthermore, visualizing the distribution of primary sources geographically can illuminate unexpected biases and patterns in historical record-keeping, adding another layer of nuance to historical interpretation. This interdisciplinary approach adds a new dimension to historical research, moving beyond simple narrative and towards a more complex and nuanced understanding.

Social scientists benefit immensely from data visualization techniques. Consider sociologists studying social networks; visualizing these relationships allows for the identification of key influencers, the detection of community structures, and the understanding of information flow within a society. Visualizations can make complex social dynamics—like the spread of misinformation or the evolution of social movements—accessible and easier to comprehend. Moreover, visualizing demographic data across various metrics allows for effective communication of social inequalities and disparities. Data visualization empowers researchers to not only identify problems but to advocate for solutions more effectively, offering compelling evidence to support their findings and making their research more impactful. Think of researchers studying crime rates, visualizing those rates on maps can clearly indicate crime hotspots, leading to more effective resource allocation and crime prevention strategies. The ability to quickly and accurately communicate complex social issues through visualization is invaluable. This allows for more effective policy recommendations based on data-driven insights rather than assumptions.

The application of visualization extends beyond traditional research contexts. In anthropology, visual ethnography allows researchers to represent and interpret cultural practices and beliefs visually. Photographs, videos, and interactive maps can be combined to create rich and nuanced representations of communities and their customs. This approach is incredibly valuable for understanding cultural diversity and promoting intercultural dialogue, moving beyond the limitations of purely textual descriptions. This is extremely valuable for breaking down stereotypes and promoting understanding between cultures. By creating visually rich representations of cultural practices, we are better equipped to challenge biases and appreciate the complexity of cultural expressions. This method contributes significantly to fostering understanding and promoting intercultural sensitivity, particularly in the academic community and beyond.

Using visualizations in the humanities doesn't just enhance understanding; it fosters innovative research questions. By visualizing the frequency of certain words or phrases in a literary cor-

pus, researchers can uncover underlying themes and patterns, enriching literary criticism and text analysis. This goes beyond simple keyword analysis; the ability to create dynamic visualizations which allows researchers to explore relationships between ideas and themes within literary works in new and exciting ways. For instance, mapping the character relationships in a novel can reveal unforeseen complexities and layers of meaning that might be missed in a traditional reading. These visual representations can be used to create new analytical approaches, fostering creative and innovative research in literary studies, offering fresh insights into long-studied texts and authors.
In conclusion, the intersection of data visualization and different academic disciplines yields incredibly rich results. By embracing an interdisciplinary approach, researchers across the spectrum can effectively leverage the power of visuals to enhance research methodologies, foster more effective communication of findings, and ultimately advance our collective knowledge. This approach opens up exciting new avenues of research and exploration, enhancing our ability to understand and interact with the world around us through the power of visualization. The future of research, particularly in the realm of big data, will certainly hinge on the integration of these multidisciplinary perspectives.

4.1 Humanities & Data

Humanities scholars, often perceived as distant from the quantitative world, are increasingly recognizing the power of data visualization. Think of analyzing the frequency of specific words in a corpus of Shakespeare's plays to uncover thematic shifts across his career, or mapping the geographical origins of historical manuscripts to trace intellectual exchanges across continents. These are not mere exercises in technical proficiency; they are deeply interpretive acts, enriching our understanding of complex historical and cultural phenomena. The data becomes a

lens, not an end in itself, illuminating subtle patterns and relationships invisible through traditional textual analysis. We begin to see how numbers, carefully collected and thoughtfully visualized, can bring new life to the humanistic inquiry, adding layers of complexity and nuance to established interpretations.
Consider the possibilities presented by visual representations of textual data. Imagine creating a network graph depicting the relationships between characters in a novel, the lines connecting them representing the frequency and nature of their interactions. Or visualizing the evolution of a particular theme or motif throughout a literary work by charting its appearance over time. Such visualizations aren't merely decorative; they transform our understanding of narrative structure, character development, and the subtle interplay of thematic elements. They offer a compelling visual narrative that complements and enhances the traditional literary analysis. This merging of humanistic interpretation with data-driven visualization is nothing short of transformative, allowing for a deeper, richer engagement with the source material.
Beyond textual analysis, data visualization offers humanistic researchers a potent tool for examining material culture. Imagine analyzing the stylistic variations in pottery shards across different time periods and regions, using geographic information systems (GIS) to create maps illustrating the spread of particular designs and techniques. Such an approach might reveal trade routes, migration patterns, and cultural influences never before apparent. It allows us to connect seemingly disparate artifacts, establishing connections across space and time, enriching the story each object tells. This application isn't about reducing the rich cultural significance to mere numbers, instead it deepens the understanding of the context, the history and the narrative around that artifact.
Furthermore, visual representations of large datasets can challenge existing narratives and uncover hidden biases within humanities research. For example, a visualization of gender repre-

sentation in historical archives might expose systematic underrepresentation of women, prompting a reassessment of existing scholarship and highlighting areas requiring further investigation. This isn't about imposing a quantitative framework onto qualitative research, instead this enhances the qualitative analysis with numerical data. The methodology provides a new perspective, unveiling potential biases embedded in both data and its interpretation. This critical self-awareness is vital for the advancement of humanities scholarship in the digital age. This is because objectivity is always an aspiration, never a given.

However, the responsible use of data within the humanities demands critical reflection. We must acknowledge the inherent limitations of data, recognizing that it's only a partial representation of reality. The selection of data, the methods of collection, and the choices made in visualization all introduce biases that must be carefully considered and accounted for. Presenting a visualization without acknowledging its limitations risks misleading the audience. Transparency and critical self-reflection are vital. Therefore we must consider the ethical and methodological implications carefully, selecting appropriate visualization methods that suit our data and research questions. Data visualization must serve to enlighten, not to obfuscate.

In conclusion, the intersection of the humanities and data presents exciting possibilities for scholarly inquiry. Data visualization tools provide humanistic researchers with unprecedented opportunities to explore complex data sets, uncovering patterns and relationships previously invisible to traditional methods. However, this integration demands a critical and responsible approach, acknowledging the inherent biases and limitations of both data and visualization techniques. By combining rigorous methodological awareness with thoughtful interpretation, humanistic researchers can use data visualization to enrich and transform their field, deepening our understanding of the human experience. The humanistic perspective offers critical insights into the interpretations of data, providing a framework for ethical considerations and thoughtful interpretation. The result is a far richer understanding of the world, its history, and its diverse cultures.

4.2 Social Sciences & Data

Social scientists, unlike those in some other fields, often grapple with data that defies simple numerical representation. Think of a survey exploring public opinion on a complex policy; the numbers alone fail to capture the nuances of individual responses, the context of their beliefs, or the interplay of socioeconomic factors. Effective visualization here requires going beyond bar charts and scatter plots. We need methods that capture the richness and ambiguity inherent in human experience.
Qualitative data analysis, specifically, offers a wealth of methods ideally suited for social science research. Consider the power of network analysis to illustrate the connections within a social group, revealing patterns of influence or collaboration previously hidden within raw interview transcripts or observational data. This might involve mapping the connections between individuals in a community based on their social interactions, revealing key players and influential subgroups. The visual representation then brings clarity to intricate social structures, allowing for deeper insights than any statistical analysis could offer alone.
Concept mapping, another powerful tool, allows researchers to visually represent complex ideas and their relationships. Imagine studying the development of a social movement. A concept map can visually organize the key themes, events, and individuals involved, providing a clear overview of the movement's evolution and internal dynamics. By using nodes and connecting lines, researchers can reveal patterns and interconnections that might not be apparent in textual data alone. The resulting map serves as a powerful communication tool, allowing others to quickly grasp the essence of a complex topic.

Narrative visualization, a relatively newer approach, integrates data with storytelling. This approach moves beyond simply representing data points; it constructs narratives that engage the audience and communicate compelling findings. Consider a study of poverty and its impact on families. A narrative visualization might weave together statistical data on income levels with individual stories and photographs, showcasing both the scope of the problem and the human cost. Such an approach brings emotional depth and context to otherwise abstract numbers, fostering a stronger impact on the reader or viewer.

However, the visual presentation of social science data carries its own set of challenges. Bias, both conscious and unconscious, can easily creep into the design and interpretation of visualizations. The careful selection of data points, the choice of color palettes, even the arrangement of elements on the page – all these choices can subtly influence the viewer's understanding. Therefore, researchers must be rigorously self-aware and transparent in their methods, clearly articulating the limitations of their data and visualizations. A robust methodology section outlining the data collection process, the choices made in visual representation, and potential sources of bias is crucial for ensuring credibility and transparency.

Furthermore, ethical considerations loom large. Protecting the privacy of participants is paramount. Anonymization techniques should be employed, particularly when dealing with sensitive data such as health records, personal finances or political affiliations. Moreover, researchers should be mindful of the potential for misinterpretation or misuse of their findings. They need to present their work in a way that is both accessible to a broad audience and avoids oversimplifying complex social issues.

The integration of social sciences and data visualization isn't just about representing findings in a visually appealing way. It's about transforming data into powerful instruments for understanding the complexities of human behavior and social phenomena. It is a means to build bridges between complex quantitative and qualitative datasets, bringing a much needed degree of insight to the narratives and connections found within social science research. This involves careful consideration of the chosen methodology, a strong emphasis on responsible data handling practices, and a commitment to ethical considerations throughout the process. This approach allows social scientists to communicate their research effectively, promoting a deeper understanding of the social world around us. This integration is transforming the way we study and understand society, and the ability to skillfully visualize social science data is becoming an essential skill for all researchers.

5 Data Ethics & Bias

Data, in its raw form, is inert. It's the potential energy waiting to be unleashed, a vast landscape of numbers and narratives yearning for expression. But this potential is double-edged. Uncritical use, especially without an awareness of inherent biases, risks not only misrepresentation but also the perpetuation of harmful stereotypes and inequalities. Responsible data visualization requires a deep understanding of ethical considerations, demanding careful attention to every step of the process, from data collection to final presentation.

Our focus here is on mitigating bias. Bias isn't a single, easily identifiable entity; rather, it's a complex, multifaceted issue that can creep into data at various stages. Think of it as a subtle distortion, a lens through which we view the world, unknowingly shaping our interpretation. This distortion can stem from the way data is collected – skewed sampling methods, for instance, where certain groups are under-represented. Or it can arise during data cleaning and processing, where unconscious decisions about what to include or exclude can subtly alter the narrative. Even in the visualization process itself, the choice of chart type, color scheme, or even the labels used can inadvertently reinforce existing prejudices.

Consider, for example, a study analyzing income disparity. If the sample population predominantly represents a certain socioeconomic background, the visualization might unintentionally minimize the financial struggles faced by other demographic groups. This isn't necessarily deliberate; it's a consequence of unintentional bias in the data acquisition and analysis. Detecting such biases requires a critical and reflexive approach. We must question our data sources, meticulously scrutinize data collection methods, and rigorously evaluate the assumptions underlying our analysis.

Furthermore, visual representations themselves can amplify existing biases. A chart with a misleading scale, for instance, can easily exaggerate or diminish differences. The use of color can also be problematic. Certain colors carry cultural connotations that can unintentionally influence the viewer's perception. For instance, the association of red with negativity in many Western cultures might lead viewers to interpret red-coded data points more negatively, irrespective of the actual data.

Therefore, acknowledging and addressing potential biases is crucial. This involves a multi-pronged approach. First, strive for diverse data sources and representative sampling methods. This ensures a more accurate and inclusive representation of the reality being depicted. Second, meticulously document all data processing steps, allowing for transparency and scrutiny. This includes detailing cleaning methods, any adjustments made to the data, and the rationale behind those choices. Third, employ diverse visualization techniques and explore alternative presentations of the data to gauge the impact of visual choices on interpretation. Fourth, involve diverse perspectives in the review process. Different backgrounds and experiences can offer invaluable insights, revealing subtle biases that might otherwise go unnoticed. Finally, always present your data with clear, factual explanations, minimizing the potential for misinterpretation.

Ultimately, ethical data visualization isn't just about avoiding misrepresentation; it's about ensuring equity and fairness in the presentation and interpretation of information. It demands self-awareness, critical thinking, and a commitment to responsible data handling. Through meticulous attention to detail and a collaborative approach, we can ensure our visualizations accurately reflect the complexities of the data and contribute to a more informed and equitable understanding of the world. The creation of data visualizations should be viewed not merely as a technical process but as a social act. It holds immense power to shape perceptions, and we should wield it responsibly, ethically, and transparently. Our goal is not just to represent data, but to communicate truthfully and fairly with the ultimate aim of fostering understanding and progress. In doing so, we not only contribute to the field of data visualization, but to society's collective knowledge and ability to effectively engage with complex issues. This responsible approach ensures that our visualizations do not reinforce harmful biases, but rather contribute to a more equitable and just understanding of the world.

5.1 Responsible Data Use

Data is powerful, capable of revealing hidden patterns and informing crucial decisions. Yet, this power necessitates responsibility. Ignoring ethical considerations when working with data can lead to misleading conclusions, reinforce existing biases, and ultimately, cause harm. Responsible data use isn't merely about following regulations; it's about cultivating a mindful and ethical approach throughout the entire data lifecycle, from collection to interpretation and visualization. This demands a conscious effort to understand the potential impact of your work and a commitment to minimizing any negative consequences. Transparency and accuracy are paramount, ensuring that your visualizations accurately reflect the data and are presented without deliberate or unintentional distortion.

Consider the source of your data. Who collected it? What were their motivations? Were there any limitations in the collection process that might introduce bias? Understanding the data's origins is crucial for interpreting it accurately. Imagine a study on income inequality that only sampled one specific demographic group. The resulting visualization, however accurate for that group, would provide a severely skewed, and potentially harmful, representation of overall income inequality. It's vital to acknowledge these limitations and contextualize your findings to avoid misleading conclusions. Responsible data visualization acknowledges the data's inherent limitations, avoiding over-generalizations and explicitly addressing any potential biases.

Moreover, the methods used to process and analyze the data should be transparent and defensible. Your audience should be able to understand how the data was cleaned, transformed, and analyzed to arrive at the final visualization. If complex statistical methods were used, explain them clearly; if assumptions were made, explicitly state them. Hiding these processes undermines the credibility of your work and prevents others from assessing the validity of your findings. Think of it like a scientific experiment; transparency in methodology allows for scrutiny and replication, crucial aspects of building trust and ensuring scientific rigor. Similarly, in data visualization, transparency builds confidence in your results and promotes a greater understanding.

Beyond technical considerations, responsible data use encompasses considerations of privacy and security. Protecting sensitive personal information is paramount. Ensure your visualizations do not inadvertently reveal individuals' identities or expose private details. Anonymization techniques can be used to protect individual privacy while still allowing for meaningful analysis. Remember that even seemingly innocuous data points can be combined to identify individuals, creating a risk of exposing confidential information. This is especially crucial when dealing with personally identifiable information (PII) and health data, which require particularly stringent protections. Careful planning and the use of appropriate anonymization methods are essential for ethical and legal compliance.

Furthermore, responsible data use extends beyond the technical and legal aspects. It requires empathy and awareness of the potential social impact of your work. Consider how your visualizations might be interpreted and used by others. Could they be misinterpreted, leading to unintended consequences? Could they perpetuate stereotypes or reinforce existing biases? By considering the broader societal implications of your work, you contribute to the responsible use of data and ensure that your visualizations support positive social outcomes. A visualization illustrating crime rates in certain neighborhoods, for example, could inadvertently reinforce negative stereotypes if not carefully contextualized with socioeconomic data. Therefore, understanding the potential consequences of how your visualization might be interpreted and considering alternative viewpoints are critical steps in responsible data visualization.

Finally, remember that continuous learning and self-reflection are key to responsible data use. Stay updated on best practices and emerging ethical considerations in your field. Seek feedback from colleagues and critically evaluate your own work. The field of data science is continually evolving, with new challenges and ethical dilemmas emerging regularly. Staying informed, participating in discussions with peers and engaging with the broader data science community are vital for maintaining a responsible and ethical approach to your work. By committing to continuous learning and self-improvement, you will enhance your ethical decision-making and produce high-quality, responsible visualizations. This proactive approach contributes to the development of a responsible data culture, benefiting both the individuals who use the data and society at large.

5.2 Bias Detection

Identifying bias isn't about finding a single "bad" data point; it's about understanding the systemic influences shaping your data. Consider the source: Was the data collected in a way that inherently favored certain groups or perspectives? Think about a survey on job satisfaction—if it's only distributed to employees who actively volunteer for it, the results might not reflect the opinions of the less engaged or those facing significant workplace challenges. The selection bias skews the overall result, painting a potentially misleading picture of reality.

Understanding the limitations of your data is crucial. No dataset is perfect, and each carries inherent limitations reflecting the methods of its creation. Perhaps the sample size is too small to be truly representative; a study on the effectiveness of a new teaching method with only ten students won't yield results generalizable to an entire school district. Or, the data might be outdated. Economic figures from ten years ago, for example, likely bear little resemblance to the current economic landscape. These limitations, though seemingly obvious, can be overlooked, leading to interpretations that are far from reality. We must always remember the context in which the data exists.

Visualizations, themselves, can be manipulative. Consider a bar chart showing the increase in a company's profits. While the data may be accurate, cleverly manipulating the y-axis can exaggerate the growth, creating a more positive (or negative) impression than the raw figures would suggest. Similarly, cherry-picking data points to highlight specific trends while ignoring others creates a biased representation. Always scrutinize the axes, scales, and the overall presentation to ensure it faithfully reflects the underlying information rather than shaping a narrative.

We need to be aware of our own biases. Our personal experiences and beliefs can unconsciously influence how we interpret data. For instance, someone with a strong belief in climate change might interpret ambiguous climate data in a way that strengthens their existing beliefs, potentially overlooking contradictory information. Similarly, a strong skepticism towards a particular phenomenon might lead to dismissal of supporting evidence. Self-awareness, open-mindedness, and a commitment to rigorous methodology are critical to avoid these pitfalls.

Addressing bias isn't about discarding data; it's about acknowledging its presence and mitigating its influence. This involves using diverse data sources, employing rigorous data collection methodologies, and engaging in careful, critical analysis. It also requires considering alternative explanations for the observed trends and exploring potential confounding factors. The goal is not to eliminate bias entirely – that's often impossible – but to understand its nature, minimize its impact, and transparently acknowledge its potential limitations in the context of the visualization. Such transparency is crucial for responsible data communication and ensures credible interpretations.

In the quest for unbiased visualizations, diverse perspectives are essential. Collaborating with individuals from different backgrounds and disciplines brings multiple viewpoints to the table and allows for a more thorough examination of potential biases. A multidisciplinary approach might unearth subtle biases that could otherwise remain unnoticed. This collaborative effort fosters a more comprehensive understanding and leads to a visualization that better reflects the complexity of the data and its underlying context. The involvement of experts in the subject matter further adds another layer of validation and quality control. Their expertise can reveal potential limitations or biases that non-experts may miss.

Finally, remember that the pursuit of unbiased visualizations is an ongoing process. No single method guarantees perfect objectivity. Continual learning, critical self-reflection, and embracing a culture of transparency are essential to navigating the complexities of data and producing meaningful and ethical visualizations. This isn't a destination, but a journey – a continuous refinement of methods and perspectives, ensuring we strive for the most accurate, responsible, and unbiased representations of information. It is through diligent effort and a commitment to integrity that we can move towards a more trustworthy and informative visualization landscape. The quest for objectivity is a constant process of checking biases, questioning interpretations and seeking alternative data sources.

5.3 Data Privacy

Protecting individual privacy in the age of big data requires a multifaceted approach. It's not simply about implementing security measures; it's about fostering a culture of responsible data handling throughout the entire data lifecycle, from collection to disposal. This necessitates a shift in mindset, moving away from a solely technology-centric perspective towards a more holistic, human-centered one. Consider the implications of unknowingly sharing sensitive personal data—financial details, medical records, even geolocation—with third-party applications or websites. The potential consequences, ranging from identity theft to discrimination, are far-reaching and underscore the critical need for robust privacy protections.
Effective data privacy hinges on informed consent. Individuals must be fully aware of what data is being collected, how it will be used, and who will have access to it. This goes beyond simply checking a box; true informed consent requires clear, concise, and easily understandable language, devoid of technical jargon. Furthermore, individuals should have the right to easily access, correct, and delete their data. The power imbalance inherent in data collection necessitates mechanisms that empower individuals to control their own personal information. This includes the ability to opt out of data collection or limit the types of data shared, empowering them to exert control over their digital foot-

print.
Data anonymization and pseudonymization techniques are crucial for safeguarding privacy while still allowing for data analysis. Anonymization aims to remove all identifying information from a dataset, making it impossible to link data back to individuals. Pseudonymization, on the other hand, replaces identifying information with pseudonyms, allowing for data analysis while preserving a degree of privacy. However, it's crucial to acknowledge that these techniques are not foolproof; sophisticated methods could potentially re-identify individuals, highlighting the need for continuous improvement and adaptation in data privacy methodologies.
Data minimization is a fundamental principle of privacy-preserving data practices. This means collecting only the data absolutely necessary for a specific purpose. Avoid collecting unnecessary information, as this reduces the risk of data breaches and minimizes the potential harm caused by unauthorized access. Remember, every piece of collected data is a potential vulnerability. Therefore, the more data you collect, the larger the attack surface and the greater the potential risk. The principle of purpose limitation is equally vital; collected data should only be used for the originally stated purpose. Any deviation requires obtaining renewed consent from the individuals involved. This prevents the misuse of personal data for unintended purposes.
Implementing strong security measures is paramount in protecting data privacy. This includes robust encryption techniques to safeguard data both in transit and at rest, employing access control mechanisms to limit who can access specific data, and establishing regular security audits to identify and address vulnerabilities. Investing in robust security infrastructure is an investment in preserving data integrity and user trust. Furthermore, rigorous employee training is crucial to ensure compliance with data privacy regulations and procedures. Human error often accounts for a significant portion of security breaches, high-

lighting the importance of well-trained personnel and effective security policies.
Staying abreast of evolving data privacy regulations and best practices is essential for responsible data handling. Regulations like GDPR in Europe and CCPA in California are continuously evolving, demanding ongoing vigilance and adaptation. Organizations must invest in compliance resources to ensure they meet legal obligations and protect user privacy. Moreover, proactive engagement with relevant stakeholders, including users, regulators, and other organizations, is crucial to fostering a shared understanding and commitment to data privacy.
Finally, fostering a culture of data privacy is crucial. This involves embedding privacy considerations into every aspect of the data lifecycle, from design to disposal. Promoting transparency, empowering individuals with control over their data, and instilling a sense of responsibility among data handlers are essential steps towards building a sustainable culture of data privacy. The ultimate aim is to strike a balance between responsible data utilization and the fundamental right to privacy, an increasingly complex and crucial endeavor in our data-driven world.

6 Building Data Fluency

Start by choosing a dataset that genuinely interests you. This isn't about completing an assignment; it's about fostering a genuine connection with the data. Passion fuels persistence, and persistence is key to developing fluency. Let's say you're fascinated by the migration patterns of monarch butterflies. Finding a publicly available dataset on their migratory routes—perhaps one detailing the number of sightings across different locations and time periods—would be an excellent starting point. Don't be afraid to explore various sources; government agencies, research institutions, and even citizen science projects often share valuable data. The key is finding data that sparks your curiosity and motivates you to delve deeper. Remember, the initial exploration is crucial; it lays the foundation for your understanding and encourages the process. Think about the kind of questions you'd want to answer about this data before you start; this way you'll always remain focused on your goals.

Next, consider the type of visualization that best represents your chosen dataset and the questions you wish to address. For the monarch butterfly migration data, a map illustrating the migration routes, possibly with the density of sightings represented by color gradients, would be intuitive and effective. Alternatively, a line graph could show the change in sightings over time

at specific locations. The choice isn't always straightforward; experimenting with different visualization methods is crucial for finding the one that best communicates your insights. If you begin with charts that best reflect what you know, this will help you to understand if these are the appropriate types of charts to represent the information that you found.
Now comes the practical application. Use data visualization software to create your visualization. There are several free and user-friendly options available, such as Tableau Public, Google Charts, or even spreadsheet programs like Excel or Google Sheets, which offer basic charting capabilities. Don't get bogged down in the technical details at first; focus on conveying your findings clearly and accurately. Remember the purpose is to make the data understandable and that the chart is a tool to achieve that. The more you practice, the more comfortable and proficient you will become in using this software. It's perfectly fine to start with simple charts and gradually move to more sophisticated ones as your skills grow.
Once your visualization is complete, critique your work. Does it accurately and effectively represent the data? Is it easy to understand? Does it convey your key findings clearly? Seek feedback from peers or instructors; a fresh perspective can identify areas for improvement that might have been missed. Revise and refine your visualization based on this feedback. This iterative process is vital for developing your critical thinking skills and for gaining a better understanding of the data's implications. In addition to this, ask yourself if there's any way to improve the visualization without losing the clarity of the meaning.
Beyond specific software, understanding the underlying principles of data visualization is crucial. Learn about different chart types—bar charts, line graphs, scatter plots, maps, and more—and their respective strengths and limitations. A well-chosen chart type will significantly impact your ability to communicate your findings effectively. For example, a bar chart is ideal for comparing discrete categories, while a line graph is bet-

ter suited for showing trends over time. Reading resources and articles about best practices in data visualization can improve your skillset substantially. Think critically about the way you can improve your next design and what elements are necessary for making the visualization clear and useful. Don't feel afraid of testing different methods!
Finally, consider expanding your horizons by exploring case studies of effective data visualizations. Many online resources showcase compelling examples across various fields. Analyzing these visualizations can help you understand how others have tackled similar challenges and used visuals to tell their data stories. This can spark new ideas and help you learn to apply different visualization techniques to suit various datasets and purposes. Consider the context of the visualization; a chart designed for a scientific publication would have different characteristics than one created for a public presentation. Pay close attention to the details, from the choice of colors and fonts to the use of labels and annotations. There's always something new to learn, and the more you see successful examples, the better you'll be at building your own. Remember that data fluency is a journey, not a destination. The key is to embrace the process of experimentation, learning, and continual refinement.

6.1 Practical Exercises

Let's start with a simple exercise: find a dataset you're interested in. It could be anything—the number of books checked out from your local library over the past year, the average temperature in your city each month, or even the number of times you've used a specific app. The key is to choose something that genuinely sparks your curiosity. Don't be afraid to choose something seemingly small; even small datasets can reveal interesting patterns. The process of choosing and understanding the dataset is just as important as creating the visualization itself.

Next, consider the type of data you've gathered. Is it quantitative, dealing with numbers and measurements, or is it qualitative, focusing on descriptions and observations? This distinction will heavily influence your choice of visualization method. For quantitative data, a bar chart might clearly show differences between categories, while a line graph could illustrate trends over time. For qualitative data, you might find a concept map or a network diagram more effective in representing relationships and ideas. Experiment to find the right tool for the job.

Now, let's move onto the actual visualization. Don't get bogged down in complex software immediately. Start with simple tools—even pen and paper can be surprisingly effective. Sketch out some ideas. Try different approaches; a scatter plot might highlight correlations, while a pie chart could display proportions. You're not aiming for perfection here; the goal is to experiment and discover which visual representation best conveys your data's story. Iterate on your initial designs, refining the visuals until you find something that's both clear and engaging.

Once you have a basic visualization, begin to think critically about its effectiveness. Is it easy to understand? Does it accurately reflect the data? Are there any potential biases or misleading aspects? Consider your audience – who are you trying to communicate with, and what level of data literacy do they possess? A visualization that's perfectly clear to a statistician might be confusing to someone with limited data experience. Refining your visualization based on these critical considerations ensures your final presentation is both effective and ethical.

Finally, share your work! Showing others your visualizations allows for valuable feedback. Constructive criticism helps to improve clarity and identify any hidden biases that you might have missed. Even simple visualizations, if well-executed and presented with a thoughtful narrative, can provide powerful insights. This interactive process of creation, critique, and revision is crucial for developing data fluency, transforming raw data into a compelling and understandable visual narrative. Remember, the journey is as important as the destination—the iterative nature of design is a vital skill to master. Practice is key; repeat this process with various datasets and visualization methods to build your confidence and expertise.

To further enhance your understanding, try incorporating different visualization techniques into a single project. For example, let's say you're analyzing data on consumer preferences for different types of coffee. You could use a bar chart to show the overall popularity of each type of coffee. A pie chart could supplement this by illustrating the proportion of each coffee type. Simultaneously, you could create a simple network diagram that links consumer demographics to their preferred coffee type—for example, linking age groups to the types of coffee they favor. By combining different visualizations, you can construct a more complete and nuanced understanding of your data.

The next step involves exploring different data visualization software. Familiarize yourself with tools like Tableau, Power BI, or even simpler options like Google Sheets or Excel. These programs offer a range of charting options, making the process of creating more advanced visualizations significantly easier and faster. You'll be amazed by the ease with which you can transform complex data into visually striking and informative charts. This will enable you to create increasingly sophisticated data presentations. Remember, the software is merely a tool; the critical thinking and design principles you've already developed remain central to the effectiveness of your visualization.

Beyond these foundational exercises, consider exploring advanced techniques. Think about how you could use interactive visualizations to involve your audience more directly, allowing them to explore data at their own pace and discover new insights. Imagine a map that highlights different regions based on the data, allowing the user to click on specific areas to reveal more detailed information. This type of dynamic interaction transforms the experience from a static visualization to an exploration of knowledge.

Remember that data visualization isn't just about making pretty pictures; it's about clear communication. Your goal is to use visuals to tell a story, to convey your key findings effectively. Always keep the audience in mind and choose visualization techniques that best suit your message and their level of understanding. Effective data visualization is a combination of technical skill and thoughtful storytelling.
Finally, don't be afraid to experiment and push your creative boundaries. Explore unconventional visualization methods, look at examples from different disciplines, and strive to create visuals that are not only informative but also visually stunning. The field of data visualization is constantly evolving, and by continuously challenging yourself and exploring new techniques, you'll unlock a deeper understanding of data and its potential to tell powerful and engaging stories. The ability to see data in a new light, to transform numbers and insights into compelling visuals, is a skill that will benefit you throughout your academic career and beyond.

6.2 Case Studies

Let's examine a case where a historical society aimed to showcase the migration patterns of a specific ethnic group over three centuries. Simple bar graphs showing population numbers in different regions wouldn't convey the nuanced story of displacement, resettlement, and community building. Instead, they chose an animated map. Each year's data appeared as a shimmering overlay, showing population density shifting across the landscape. This created a compelling visual narrative of the group's journey, far exceeding the impact of static charts. The choice of animation, a dynamic visual approach, highlighted the temporal dimension crucial to understanding these migration patterns. This demonstrates the power of selecting the right visualization type to best represent the data's inherent characteristics.

Consider a contrasting example: a public health organization needed to communicate the complex interplay of various risk factors contributing to a particular disease. A static infographic might have listed factors, but it wouldn't convey the intricate relationships between them. They opted for a network graph, where each node represented a risk factor, and the connecting lines indicated the strength of their influence on each other. This allowed viewers to immediately grasp the interconnectedness of factors, something lost in a simple list or even a series of bar charts. This visualization helped them highlight critical correlations, informing intervention strategies far more effectively than other visualization approaches could have.

A third scenario highlights the use of visualization in a business context. A marketing team wanted to show how customer preferences changed across several product lines over five years. Simply presenting sales figures wouldn't be effective. They used a series of clustered bar charts, each cluster representing a product line. The bars within each cluster tracked sales over the years, revealing trends and shifts in popularity. This effectively illustrated product life cycles and informed future marketing strategies. Choosing this format allowed a comparison across different products, simplifying complex datasets into a clear and understandable format that facilitated decision-making.

Now, consider a study exploring the spread of a new social media trend across different demographic groups. A traditional graph displaying numerical data wouldn't fully capture the organic, unpredictable nature of such trends. Researchers instead employed a visually rich network analysis, mapping connections between users as they adopted the trend. The visualization highlighted how influence spread through specific communities and how certain individuals served as pivotal "hubs" within the network. Such a method beautifully illustrates complex data relationships, providing insights that numbers alone couldn't. This type of rich visualization allowed for the analysis of qualitative elements within a quantitative dataset, producing greater nu-

ance in the results.
Moving beyond quantitative data, consider a team of anthropologists documenting the daily lives of a remote community. Photographs, videos, and field notes—rich qualitative data—needed effective organization and presentation. They built a multimedia timeline, integrating images, audio clips, and written observations, thus illustrating the daily rhythms and cultural practices. This approach wasn't simply about representing the data but about crafting a narrative that respected the community's stories and provided deeper cultural understanding. This method emphasized storytelling with the data, going beyond the simple summary and providing a richer experience.
Finally, imagine a city planning department seeking to understand the effect of a new transportation system on traffic patterns. Traditional maps alone weren't enough. They combined real-time traffic data with heatmaps to visualize congestion levels before and after system implementation. The resulting dynamic visualization offered a compelling demonstration of the system's impact, allowing policymakers to see the effects in a tangible and readily comprehensible format. This sophisticated visualization method allowed them to not only understand the data but also to communicate the findings effectively to stakeholders and the public. The heatmap, in conjunction with the traffic data, demonstrated the efficacy of the visualization methodology utilized. The resulting visualization was intuitive and easily understood.

6.3 Further Exploration

Having grasped the fundamentals of data visualization, it's time to delve deeper, experimenting with more complex visualizations and exploring less conventional approaches. Consider combining different visualization techniques within a single project; a geographical map overlaid with network analysis, for instance, could reveal fascinating patterns in migration or disease spread. This synergistic approach offers richer insights than using any single method in isolation. Don't be afraid to iterate, refining your visual representations based on feedback and further analysis. Remember, the goal is to communicate effectively, and that sometimes requires more than one attempt.

Now, let's address the crucial aspect of interactive visualizations. Static images, while useful, often lack the depth and engagement offered by interactive tools. Software like Tableau or D3.js enable the creation of dynamic dashboards, allowing users to explore data at their own pace, filtering, zooming, and uncovering hidden details. Learning even basic interactive coding can significantly enhance your ability to create compelling and informative visualizations, empowering you to tailor the user experience.

Beyond the technical aspects, consider the narrative arc of your visualization. Just as a well-written story needs a beginning, middle, and end, so too should your data representation. Lead the viewer through a journey of discovery, gradually revealing information and building towards a clear conclusion. Think carefully about the order in which you present your data; the narrative structure is pivotal for effective communication. A poorly structured visualization, even if technically sound, can leave your audience confused and disengaged.

Another critical area often overlooked is the aesthetic dimension. A beautiful visualization is not just pleasing to the eye; it also enhances clarity and memorability. Pay close attention to color palettes, typography, and overall layout. The use of whitespace is equally crucial; it provides visual breathing room and prevents the visualization from becoming cluttered. Explore design principles, studying the work of renowned data visualization artists to understand how aesthetic choices can impact communication. Remember, visual appeal can significantly amplify the impact of your message.

Moving beyond individual projects, consider the creation of a visual portfolio showcasing your work. This portfolio will serve as a tangible demonstration of your skills and knowledge, a testament to your journey through the world of data visualization. This isn't just about compiling your projects; it's about curating a collection that highlights your growth, experimentation, and unique approach to visual communication. A well-structured portfolio is a powerful tool, particularly when applying for jobs or seeking further educational opportunities.

Finally, embrace the continuous learning inherent in this field. Data visualization is a dynamic field constantly evolving with technological advancements and emerging best practices. Stay updated on the latest tools, techniques, and trends by reading relevant publications, attending conferences, and engaging with the online data visualization community. Follow influential figures, participate in online discussions, and actively seek feedback on your work. Continuous learning is not an optional extra; it is the key to remaining relevant and competitive in this rapidly evolving field. Through persistent practice, mindful exploration, and a commitment to continuous improvement, you will unlock the true potential of data visualization. Remember, the journey of mastery is a marathon, not a sprint.

7 The Future of Data Visualization

Augmented reality (AR) overlays digital information onto the real world, offering interactive and immersive data experiences. Imagine exploring a 3D model of a city's infrastructure, with real-time traffic flow data superimposed on the buildings. This technology allows for a deeper understanding of complex datasets by placing them directly within their context, enhancing engagement and comprehension far beyond traditional static visualizations. The potential applications extend across numerous fields, from urban planning and environmental monitoring to medical diagnostics and educational simulations. AR's capacity to seamlessly integrate data into our daily environment promises to revolutionize how we interact with and interpret information.
The increasing availability of vast, complex datasets, often described as "big data," demands new visualization approaches. Traditional methods struggle to effectively represent the intricate relationships and patterns inherent in such massive quantities of information. Consequently, researchers are exploring innovative techniques like dimensionality reduction and interactive exploration tools that enable users to navigate and understand these datasets more effectively. These advances are crucial for making sense of the overwhelming amount of data gener-

ated daily across various sectors, facilitating informed decision-making and fostering new discoveries. The focus shifts from simply presenting data to empowering users to actively engage with it and uncover meaningful insights.
Artificial intelligence (AI) and machine learning (ML) algorithms are rapidly transforming data visualization. AI can automate tasks like data cleaning, feature extraction, and the selection of optimal visualization methods, significantly streamlining the process. Moreover, AI-powered systems can identify patterns and insights within data that might be missed by human analysts, leading to new discoveries and more accurate interpretations. This synergy between human expertise and AI capabilities is not about replacing human analysts, but rather augmenting their abilities, allowing them to focus on higher-level tasks like interpretation and decision-making. The future will likely involve sophisticated hybrid systems where human judgment complements the analytical power of AI.
The integration of data visualization with other technologies, such as the Internet of Things (IoT) and blockchain, is opening up new possibilities. IoT devices constantly generate massive streams of real-time data, demanding dynamic and responsive visualizations capable of updating in real time. Blockchain technology, with its secure and transparent data storage capabilities, can enhance the credibility and trustworthiness of visualizations by providing an immutable record of the data sources and analysis processes. The combination of these technologies promises to create entirely new categories of data-driven applications, impacting diverse sectors from supply chain management to healthcare.
Ethical considerations will become increasingly prominent in data visualization. As the use of data visualization expands, addressing issues of bias, privacy, and misrepresentation is crucial. Transparency in data sources and methodologies is paramount to ensuring accountability and avoiding the potential for manipulation. Developing tools and techniques to detect and mitigate

biases within both data and visualization choices will be critical for promoting fairness and responsible data use. Educational initiatives focusing on data literacy and critical thinking skills are essential to equip users with the ability to interpret visualizations effectively and critically evaluate their potential biases. The future of data visualization hinges on our ability to use it responsibly and ethically.

The future of data visualization is not merely about better technology; it's about better communication. The ultimate goal is to make complex information accessible and understandable to a broader audience, facilitating data-driven decision-making across all sectors of society. This involves simplifying interfaces, using intuitive visual representations, and designing visualizations that cater to different learning styles and levels of expertise. The emphasis will increasingly shift towards user-centered design, making sure visualizations are not only visually appealing but also effective in conveying their intended message clearly and unambiguously. Effective communication of complex data, not just its presentation, will define successful visualizations.

Data visualization is evolving beyond static representations toward dynamic and interactive experiences. This trend moves towards a more active and engaged audience, fostering critical thinking and deeper engagement with data. Interactive dashboards and visualizations that allow users to explore datasets, filter information, and customize their view enhance understanding and knowledge retention. Moreover, these dynamic approaches encourage exploration and discovery, allowing users to uncover hidden patterns and insights that static visualizations might miss. This shift towards interactive visualizations will create more engaging and effective ways to communicate insights to diverse audiences and facilitate knowledge discovery.

7.1 Emerging Technologies

Virtual and augmented reality are transforming how we interact with data. Imagine exploring a complex dataset not through static charts, but by virtually stepping inside a 3D model of your data, manipulating variables in real-time, and gaining intuitive insights previously hidden within spreadsheets. Augmented reality overlays data visualizations onto the real world, enabling on-site analysis and decision-making. For example, construction workers could use AR glasses to see blueprints and project models overlaid on the actual site, streamlining workflow and minimizing errors. This immersive approach fosters a deeper understanding and encourages more intuitive problem-solving. The possibilities extend far beyond these examples; consider using VR to explore intricate social networks or AR to analyze real-time sensor data in environmental monitoring. These technologies are rapidly maturing, promising a more engaging and effective way to interact with information.
The integration of artificial intelligence offers exciting possibilities for automation and analysis. AI algorithms can automate the process of data cleaning, pre-processing, and even visualization selection, reducing the time and expertise needed to create compelling visuals. Furthermore, AI can assist in identifying patterns and insights within datasets that might be missed by human eyes. This includes anomaly detection, predictive modeling, and generating interactive visualizations that adapt to the user's needs. However, it's crucial to remember that AI is a tool, and its outputs require careful review and interpretation. Overreliance on AI without critical assessment could lead to biases or misleading interpretations. The responsible use of AI in data visualization involves careful consideration of its limitations and a focus on human-in-the-loop approaches, combining the power of AI with human judgment and expertise.
Machine learning is changing how we create and interpret data

visualizations. By training algorithms on large datasets of visualizations, we can develop systems that automatically generate visualizations suitable for a given dataset and purpose. Imagine a system that, given a dataset, automatically selects the most appropriate chart type, optimizes its design, and even generates a compelling narrative to accompany it. This would democratize data visualization, making it accessible to a broader audience. However, it's essential to address the potential for bias in these algorithms. If the training data reflects existing biases, the generated visualizations will likely perpetuate those biases. Addressing this bias is crucial to ensure fairness and equitable outcomes. Machine learning's potential to streamline and personalize the data visualization process is undeniable, but careful attention must be given to ethical considerations.

Beyond these, several emerging technologies are subtly but significantly influencing the field. For example, advancements in 3D printing allow for the creation of tangible, physical representations of data—offering an alternative way to comprehend complex relationships and share information with a wider audience. Imagine using a 3D-printed model to represent a network of relationships, where the size and position of elements reflect their influence, making the network's structure easily and visually accessible. Similarly, advances in haptics—the science of touch—promise a more immersive data experience, allowing users to feel data through tangible feedback. These developments are broadening the range of ways we can interact with and understand data, particularly useful for visually impaired users or those who learn best through hands-on experiences. They offer a future where data visualization is no longer limited to screens and digital interfaces.

Finally, it's important to consider the broader societal impact of these technological advances. As data visualization becomes more accessible and powerful, it's essential to ensure responsible use and ethical considerations remain at the forefront. The potential for misinformation and manipulation is significant; therefore, critical thinking and media literacy skills are paramount. The development of tools and methods for evaluating the credibility and objectivity of visualizations is necessary. Promoting data literacy and fostering a critical understanding of data visualization is crucial in navigating this increasingly data-driven world, preventing misuse, and promoting informed decision-making across all sectors of society. The future of data visualization isn't just about technological advancement; it's about responsible innovation that benefits humanity as a whole.

7.2 Trends & Predictions

Interactive data storytelling will become increasingly prevalent. Imagine exploring complex datasets not just through static charts, but through immersive narratives that adapt to user choices and allow for deeper, more personalized understanding. Think of it as a choose-your-own-adventure, but with data at its core. This shift requires a move beyond simple visualization towards dynamic, engaging interfaces that encourage exploration and discovery. The creation of these interactive stories will demand a new breed of data visualization specialists, comfortable with both data analysis and narrative design. Tools will emerge to facilitate this, simplifying the creation of sophisticated interactive experiences without demanding advanced coding skills. This democratization of interactive data storytelling will be a key trend, putting the power of data narrative into more hands than ever before.

The demand for specialized visualization for specific fields will explode. While general-purpose tools will continue to thrive, we'll see a surge in visualization techniques specifically tailored to individual disciplines. For example, medical imaging will move beyond simple scans to incorporate complex 3D models and simulations that assist diagnosis and treatment planning. Financial analysts will utilize highly sophisticated visualizations that map complex relationships between financial instruments, predicting market fluctuations with greater accuracy. Even the humanities will benefit; imagine text analysis tools visualizing the evolution of language through interactive maps showcasing word frequency and usage shifts across centuries. This specialization will require collaboration between data scientists and domain experts, blending technical expertise with in-depth subject matter knowledge.

Ethical considerations will become even more central to data visualization. As data visualization tools become more powerful, the potential for misuse increases. This means that the focus on responsible data use, bias detection, and data privacy will only intensify. Expect to see increased emphasis on algorithmic transparency, methods for visualizing bias, and the development of ethical guidelines for data visualization professionals. The concept of "data provenance" – the ability to trace the origin and history of data – will become critically important in maintaining data integrity and trust. This will necessitate the development of new educational programs and certifications focused on ethical data practices within the visualization process. We will see greater regulatory scrutiny of data visualization practices, particularly in sensitive sectors like healthcare and finance.

Automation will change the data visualization landscape. While many fear job displacement, the reality is likely to be more nuanced. Automation will handle the more repetitive tasks of data cleaning, formatting, and even generating basic visualizations. This frees up human analysts to focus on the more creative and strategic aspects of data visualization, such as designing compelling narratives, interpreting complex results, and communicating findings effectively. The shift will be toward human-machine collaboration, where humans guide and refine the automated processes, resulting in enhanced efficiency and accuracy. This change demands a workforce adaptable to new technologies and willing to embrace automation as a tool rather than a threat. The focus will shift towards higher-level skills: critical

thinking, problem-solving, and communication.
Data visualization will become increasingly integrated into everyday life. We are already seeing the beginnings of this trend with the growing use of data dashboards in homes to monitor energy consumption and health trackers providing personalized insights. In the coming years, this integration will deepen, as data visualization becomes an essential tool in many aspects of our lives—from personal finance management to city planning and environmental monitoring. The challenge will be to ensure that this integration is seamless and intuitive, without overwhelming individuals with data overload. The design of user-friendly and accessible interfaces will be paramount. This requires careful consideration of user experience and the development of tools that are both informative and engaging, regardless of technical expertise.

www.ingramcontent.com/pod-product-compliance
Ingram Content Group UK Ltd.
Pitfield, Milton Keynes, MK11 3LW, UK
UKHW020155290926
15471UKWH00005B/891

9 798349 393143